S0-BUD-888

IF YOU LIVED ON MARS

IF YOU LIVED ON MARS

MELVIN BERGER

LODESTAR BOOKS
E. P. Dutton New York

Unless otherwise credited, all photographs
in this book courtesy of NASA

Library of Congress Cataloging-in-Publication Data

Berger, Melvin.
If you lived on Mars / by Melvin Berger.
p. cm.
Bibliography: p.
Includes index.
Summary: Postulates what life would be like living in a colony on Mars and presents up-to-date scientific knowledge of that planet.
ISBN 0-525-67260-5
1. Mars (Planet)—Juvenile literature. 2. Life on other planets—Juvenile literature. [1. Mars (Planet) 2. Life on other planets. 3. Outer space—Exploration.] I. Title.
QB641.B46 1988 89-9105
919.9'2304—dc19 CIP
AC

Published in the United States by
E. P. Dutton, New York, N.Y.,
a division of NAL Penguin Inc.

Published simultaneously in Canada by
Fitzhenry & Whiteside Limited, Toronto

Editor: Virginia Buckley Designer: Robin Malkin

Printed in the U.S.A. First Edition
10 9 8 7 6 5 4 3 2 1

CONTENTS

An eight-page color insert begins on page 27.

1

THE COLONY

IF YOU LIVED ON MARS . . . your home would be inside the colony.

The colony is the only place on the entire planet where people can live. In fact, it is the only place on Mars where there is any life at all.

The colony looks just like a covered shopping mall on Earth, except that it is much bigger and and is built partly underground and partly above ground. The Mars colony covers nearly ten square miles. And in the central area, the roof is high enough to enclose buildings five stories tall!

Under the huge colony roof are all the buildings you would find in a self-contained small town on Earth. There is housing for the population, which numbers about 1,600. There are stores selling all the various products that people need. There are factories turning out manufactured goods. There are offices that serve all the businesses on Mars. There are also schools, from nursery school through high school. Students must go to Earth for their college education.

There is one very small hospital. Before the settlers arrived, there was no life on Mars. Therefore, the only germs on Mars are those they brought with them. As a result, there are very few germ-caused diseases on Mars. Most of the Mars residents are very healthy and seldom get sick. However, the hospital is equipped to handle the full range of illnesses. It is just too long a trip to send patients back to Earth for treatment.

Attached to one end of the colony are greenhouses for growing food. A spaceport for launching and landing rockets and airships has been built at the other end. Some distance from the colony are the mines, where a number of settlers work.

In short, everything that people need to live fully and well is found in or near the colony.

Even though the colony is spread out over several square miles, you almost never lose your way. That is because a grid of parallel streets runs through the settlement. The streets look like a cross between city streets on Earth and the aisles of a supermarket. Vehicles travel in the middle of the street. People walk along the sides.

Small open carts provide transportation. The carts are roofless because the climate within the colony is completely controlled. It never rains or snows; neither does it get too hot or too cold.

The carts are completely automatic. Each one has a computer keyboard. You punch in the code for your destination. The computer then directs the cart to follow a pattern of wires beneath the roadways. You get where you are going by the shortest, fastest route.

You don't have to do a thing. The cart even has a radar setup that immediately stops it if something is in its way. You can just sit back and enjoy the scenery!

Passenger carts share the roadways with a small fleet of freight carts. These vehicles look like trucks. They move heavy objects and goods from place to place.

The houses people live in, as well as the other buildings inside the colony, are made of just two materials—concrete and metal. Trees do not naturally grow on Mars. This makes wood extremely rare. Almost all of it is brought up from Earth in spaceships at great expense and is very costly.

Also, there are no oil wells on Mars. Hence, there is no plastic, since plastic is made from oil. A few plastic items are rocketed up from Earth. But they cost as much as the finest china.

Workers use the Martian soil to make the concrete for building construction. The soil is red in color. This gives all the buildings a pink-gray tinge. Otherwise the structures on Mars look much like the concrete buildings on Earth, with fewer windows. (In part, this is because it is difficult and expensive to make glass on Mars or bring it up from Earth. In part, it is because without fresh, outdoor air and bright sunlight, there is little to be gained from windows.) Also, the outer walls are much thinner, since they do not need to protect people from the weather.

The metal used in construction also comes from the

soil of Mars. Iron is the most abundant metal on this planet. About 12 percent of the soil is made up of various compounds of iron. Other metals in the soil are aluminum, 8 percent; titanium, 4 percent; and magnesium, 2 percent.

Scientists have found a cheap, easy way to separate out the metals from the soil. Basically, the method involves digging up soil from outside the colony, placing it in huge vats, and sending a powerful electric current through the vats. The pure metals are then collected and worked into the forms that are needed.

The colony itself is not built on the surface of Mars. Rather, it is dug into the Martian soil. Most of the colony is below ground level. In addition, the roof is covered with a thick layer of soil. The soil acts as a shield. It protects the people in the colony from the dangerous radiation of space. This radiation includes ultraviolet rays, cosmic rays, and even some X rays.

The same radiation from outer space threatens people on Earth. But the Earth's atmosphere shields them from the dangerous rays. Not so the atmosphere on Mars. It is too thin to keep out the rays.

Inside the colony, though, you are safe and comfortable. And you feel happy to be a pioneer on this exciting planet!

YOUR HOME ON MARS

IF YOU LIVED ON MARS . . . your house would be in a five-story underground building. The building would have many separate apartments.

There are no private houses on Mars. The reason is simple: There is just not enough space inside the colony. Land is too valuable to be taken up with individual, small structures.

Since the colony is limited in size, everything needs to be very compact. Thus, all the apartments are quite small. The plan is to let as many people as possible live comfortably within the boundaries of the colony.

Much of the furniture is actually built into the walls of each apartment. Your bedroom, for example, has concrete walls, floor, and ceiling. The bed, dresser, and desk are made of metal. But they are not separate pieces of furniture. They are part of the room itself, attached to the walls and floor. When you wake up in the morning and sit on the edge of your bed, you can choose your clothes, dress, and pack your schoolbooks—all without taking one step!

The clothes that hang in your closet look much like the ones people wear on Earth. But the fabrics feel different. Few articles of clothing on Mars are made of cotton, wool, or any other natural material. The reason is that no cotton plants grow on Mars from which to make cotton thread. Nor are enough sheep, cattle, or pigs raised here for wool and leather. In addition, not having oil wells means there are no raw materials for synthetic fabrics such as nylon and polyester.

The fact is, almost all of your clothes are made of metal. The metal is formed into very long, thin, flexible threads. Workers then treat the surface of the metal threads to make them feel like cotton, silk, or wool. And when these metal threads are woven into cloth, the result is remarkably like natural or synthetic fabrics on Earth.

The first people who lived on Mars, starting in the year 2027, had to manage with the small amounts of building supplies and fabric that they brought up with them in spaceships. Then, in the early years, people figured out ways to make similar materials out of metal.

Now, however, fifty-six years have passed. People are beginning to bring trees, plants, and animals to Mars, to be grown and raised in the colony. And as trips between Earth and Mars become more frequent and cheaper, shipments of oil to make plastics and synthetic fabrics are also being sent up from Earth.

Your children will probably have a very wide choice of fabrics, natural and manmade, from Mars or from Earth.

In your apartment, meals are cooked on an electric stove or in a microwave oven. There are no gas stoves on Mars. Electricity is the only source of power. It is used for cooking, heating, and for light.

Since the roof of the entire colony is covered with soil, there is no natural light. Artificial lights in the streets are on almost all the time. Only during the night hours are some lights shut off or dimmed.

The big advantage of using electricity on Mars is its low cost. Most of the energy comes from the panels of the huge solar cells placed on the soil that covers the colony. The other source is an electric generating plant that is run by nuclear fusion. This advanced process releases immense amounts of heat. The heat is then used to power the generator that produces the electricity.

Water, on the other hand, is very expensive. Right now there is no natural water on the planet, and all of it must be produced in a factory. Because water is very scarce, something as simple as washing up in the morning is a costly proposition. One day, scientists hope to have a natural water supply—enough for everyone to take long showers and to wash their dishes under running water. But right now you, like everyone else on Mars, must use as little water as possible.

The computer in your living room is as common on Mars as the bed in your bedroom. Everyone has one. In fact, some of your friends have two or three, one for each member of the family.

You use your computer as a telephone, or rather as a picturephone. To speak to a friend, you enter his or her number on the keyboard. When the friend answers, you see the person on your computer screen and your friend sees you. Then, both of you talk to each other as though you were in the same room.

All of your home entertainment equipment works through the computer too. The computer monitor serves as a television screen. Every apartment in the colony has cable TV with a choice of five channels.

A number of different types of programs are shown on Mars TV. Some are popular shows from the United States. But the local news and a growing number of comedies, dramas, and music shows are produced in studios on Mars.

Since the Mars colony is a joint project with the Soviet Union, there are also quite a few Russian families living in the colony. One channel, therefore, is devoted only to Russian-language programs. As with the broadcasts in English, some come from Earth and some are filmed in the colony.

No one on Mars plays audio- or videotapes, records, or even compact discs. Instead they use tiny computer chips. The chips—about the size of a postage stamp—are inserted into the computer, which "plays" them back. Some chips are recordings of music. Others have both sound and pictures and carry entire movies or TV shows.

The founders of the colony chose computer chips for home recreation as a way of saving space—both in the cargo rocketed up from Earth and in the small apartments on Mars.

Just as on Earth, the same computer serves several other functions as well. Suppose you're doing your homework and need some facts or figures. You use your computer to get the information from a data bank. Or, imagine that your parents have work that they were not able to finish at their offices. They use the computer to get in touch with their files and do the work at home.

When your folks want to do the food shopping they call up a list of the items and prices at the supermarket. Then they punch in the ones they want and the order is delivered within an hour. If they want tickets to a concert or tennis match, or if they need to find out what's playing at the movie house, they also use the computer. And of course there are always plenty of computer games for the whole family to enjoy.

Life on Mars is not all fun and games, however. Conditions are far more primitive than on Earth. But what can you expect? The colony is in its infancy. Living in space is still at a very early stage.

Yours is the first space colony on a planet. Only the colony on the Moon—established in the year 2017, ten years before the Mars colony—is older. There's still much to learn before the colony becomes as complete as a small city on Earth. But in the meantime, experts are carefully studying every aspect of life in the colony both to help them improve conditions and to plan possible new colonies on other planets.

3

THE HISTORY OF THE COLONY

If you lived on Mars . . . you would go to school like young people on Earth.

Your school building looks much like the school buildings on Earth, except that it is very much smaller than most of them. The subjects you study are the same as those taught on Earth. In addition, though, everyone learns the history of Mars.

You learn that for many hundreds of years astronomers on Earth looked at Mars through their telescopes. A good number thought they saw signs of life and civilization on what they called the Red Planet.

In 1877 an Italian astronomer, Giovanni Schiaparelli, was studying Mars through his telescope. He saw about forty thin, dark lines crisscrossing the surface, which he called *canali*.

Canali is an Italian word meaning "channels." Channels are natural waterways, like the English

Channel. But someone translated *canali* as "canals." Canals are manmade waterways, like the Erie Canal. Out of this mix-up came the idea that there were creatures on Mars who dug canals.

Many other astronomers also reported seeing lines on Mars. Among them was the American astronomer Percival Lowell. Lowell also found several dark places where the canals crossed. He decided that these were oases. They looked dark, he wrote, because of the trees and houses there. The *canali* had been dug by Martians, Lowell concluded, to bring water from the oases to the dry parts of the planet.

Lowell was an outstanding astronomer. Others accepted his conclusions. And for nearly a hundred years everyone thought there were canals dug by Martians on this planet.

Not much more was learned about Mars until the 1960s. That was when some scientists made up their minds to put a human on Mars. To get ready for a landing and to learn more about the planet, the United States launched the unmanned spaceship *Mariner 4* in 1965. It was aimed to fly by Mars. As it passed, it took pictures and radioed them down to Earth. The photos gave scientists their first close-up view of the Red Planet.

Four years later two more space shots, *Mariner 6* and *Mariner 7,* were launched. They came even closer to Mars. Their photos were better, too, showing more details.

The *Mariner 9* mission in 1971 was really special. It didn't just fly past Mars like the others. *Mariner 9* actually went into orbit around the planet. It flew

Mariner 7 took some excellent photographs of Mars.

within 800 miles of the surface. And for nearly a year it continued to circle Mars. Television cameras sent a stream of pictures back to Earth. Other devices on board measured the planet's heat and light.

Mariner 9 also took pictures of places where the lines had been seen by Lowell and others. But none of these shots showed canals. Instead, they showed hundreds and hundreds of very large rocks. When the sun was low, each boulder cast a shadow. This created an optical illusion that looked like many long, solid lines. It was these lines that earlier scientists had seen and imagined were the *canali.*

A giant step forward in the exploration of Mars came in 1976. NASA (the National Aeronautics and Space Administration) landed two Viking probes on the surface of Mars. *Viking 1* arrived on July 20, *Viking 2* on September 3. The two landing spots were 1,800 miles apart. This way scientists could study two completely separate areas.

The Viking instruments sent back hundreds of close-up photos of Mars. They also reported on Mars's temperature, wind, and radiation. And, most exciting of all, they sought an answer to the question that was on everybody's mind: Is there life on Mars?

You might think that the scientists of the Viking missions looked for big creatures, like animals or people. Not at all. They realized that such signs of life might be too hard to find. Instead they searched for germs and other tiny microbes. Such living things are spread out everywhere and in very great numbers. They can be found in the soil and in the air. And they can survive and multiply under all sorts of conditions.

In the first experiment, the Vikings were directed to scoop up samples of Martian soil and heat it. The researchers believed that if there were living things in the soil, they would give off gases when heated. But when the Viking instruments tested the soil for such gases, they found nothing.

The next test was based on the idea that microbes "eat" some sort of food and produce some sort of waste product. In the so-called chicken soup experiment, the scientists ordered the Vikings to add liquid nutrients to the soil. Then the Viking instruments studied the air above the soil for signs of waste products.

The arm of Viking scooped up samples of soil to test for life.

The figures showed fifteen times more oxygen in the Martian air than other tests had shown was normally there.

At first the scientists were sure this proved that there was life on Mars. But later they changed their minds. They realized that the nutrients contained water. Water is not normally in Martian soil. The water caused a chemical reaction, and the reaction released the oxygen. It did not come from living microbes.

In another test the Vikings used nutrients that included radioactive atoms. These atoms give off rays and particles that can be traced. Any microbes present would take in the nutrients with the radioactive atoms. And the radioactivity would then appear in their waste products.

The scientists found a big jump in the radioactivity in the soil.

To make certain, the scientists decided to repeat the test. This time the radioactivity did not go up. This result convinced the experimenters that it was another chemical reaction, not a sign of living things.

For the final test, the scientists directed the Vikings to release carbon dioxide gas with radioactive atoms over a sample of soil. Any living microbes in the soil would take in some radioactive atoms. Then the Viking instruments tested the soil for radioactivity.

Once more the scientists noted an increase in radioactivity.

So the soil was heated to 347 degrees Fahrenheit (°F) to kill any microbes. The test was repeated. And again the radioactivity shot up.

Finally, someone figured out the reason. Ammonia had dripped into the ground from the Viking lander. And it was the ammonia that had captured the radioactive atoms in the soil.

In summary, then, all four tests seemed to show life at first. But, in the end, there was no real evidence of living things. Also, the many hundreds of pictures of Mars did not reveal one footprint, trail, burrow, or any other sign of life.

After the Viking landings, planetary exploration stopped. During the 1980s, the U.S. government made big cuts in its space budget. The loss of the shuttle *Challenger* in January 1986 also set the space program back.

Then, slowly, the pace began to pick up. Scientists launched several more unmanned missions to Mars. They collected more information on the weather, soil, and best landing spots for humans.

In 1998 the first manned spacecraft landed on Mars. Two astronauts spent four hours on the surface of Mars. They also left behind some automatic measuring devices to continue radioing data back to Earth.

Then the president of the United States got interested in Mars. He promised a permanent colony on Mars by the year 2030. Four more manned spacecraft

In the late 1990s an unmanned rover was sent to Mars to collect soil samples and take them back to Earth.

landed on Mars, each one staying longer than the one before. The last one, in 2026, carried a group of three astronauts who remained on Mars for sixteen days.

Then, in September 2027, nine astronauts and scientists came to Mars to stay. Six of them were Americans. The other three were Russians. And in the years since they landed, the colony has grown to about 1,600 inhabitants.

Of course, the nine original astronauts did not give rise to all these descendants! Most of the present population emigrated to Mars from Earth.

Mars is an average of nearly 50 million miles from Earth. Depending on the date of departure, the type of spacecraft used, and the route, the trip can take anywhere from six months to a year. But more and more people are making the journey to live in this exciting outpost of humanity.

WEAK GRAVITY AND STRONG MUSCLES

IF YOU LIVED ON MARS . . . you would find that you *must* exercise daily because the gravity on Mars is very weak.

Actually, each planet in the solar system has a different gravity, or amount of pull. The strength of the gravity mostly depends on the planet's size. The larger and heavier the planet, the greater the pull of its gravity.

The diameter of Earth at the equator, for example, is 7,927 miles. The diameter of Mars at its equator, though, is only 4,217 miles. In other words, Mars is just about half the size of Earth. Since Mars is smaller, its gravity is less. As a matter of fact, the gravity on Mars is only 38 percent as strong as the gravity on Earth. If you weighed 100 pounds on Earth you would weigh only 38 pounds on Mars.

The weaker pull on your body makes you feel light

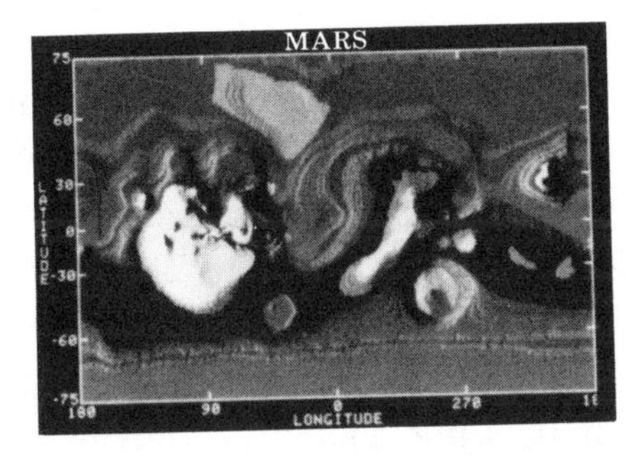

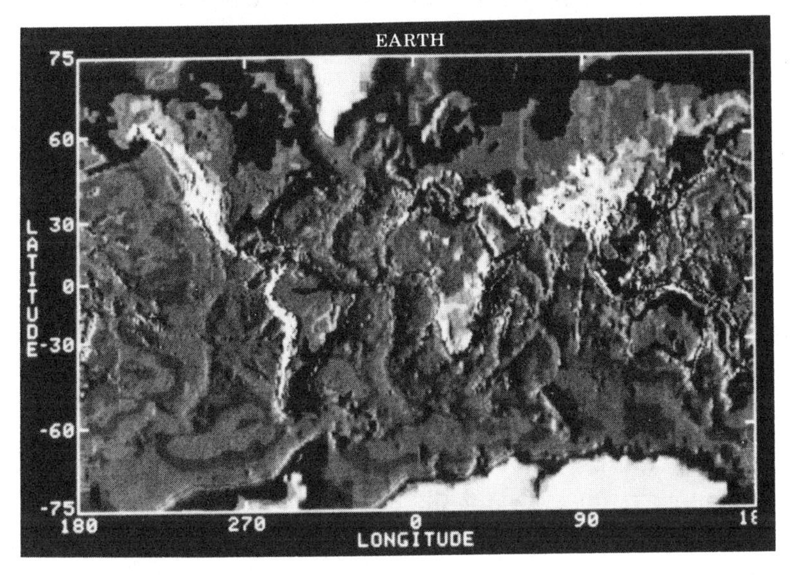

Two maps using the same radar resolution as that achieved by the Pioneer-Venus spacecraft show that Mars is only half the size of Earth.

and bouncy. Every step you take is about twice as long as on Earth. You can run much faster. And you can jump much higher and much farther. In fact, you could probably break many of the Olympic records set on Earth!

But the lower pull of gravity does not make all of your life a breeze. It also creates problems. Without the continuous push and pull against a strong gravity, your muscles become weak and small. Muscles need to work to keep up their size and strength. Otherwise they become flabby and start to waste away.

Daily exercise to keep up muscle tone is a necessity on Mars. But it is not enough to pull or push against gravity, as in many Earth exercises. Knee bends and sit-ups do not help your muscles very much on Mars because there is not enough resistance to overcome. Neither does weight lifting do your body much good. A 500-pound set of weights used for weight lifting on Earth only weighs 190 pounds on Mars.

Like everyone on Mars, you mostly use exercise machines to keep fit. You pit your muscles against springs that have the same tension, no matter what the gravity. Pushing, pulling, or twisting the springs always requires the same amount of muscular effort.

Wrestling works the same way. The exercise comes from straining your muscles against the force of your opponent's muscles, not his or her weight. Also, pitting your own muscles against each other is another approach to muscle building.

You get many chances to play different sports on Mars. But they are not the same ones as people play on Earth. Just imagine what would happen if you hit a baseball in the weak gravity on Mars! Or passed a football! Or shot a basket! Or kicked the ball in soccer! Because of the weak gravity, the ball would just keep going and going. There is no ball heavy enough or stadium or ball field large enough to hold it in.

Three sports have been found to work quite well in the weak gravity of Mars: tennis, Ping-Pong, and handball. Heavier, less bouncy balls and somewhat longer courts or tables let you play very fast-moving, exciting games.

Because of the weaker gravity, you find that you have a better posture than many people on Earth. You are also able to take deeper breaths and make better use of your rib cage, letting it expand more when you inhale.

The weaker gravity also means your body doesn't have to work so hard as you go about your daily tasks. It uses the oxygen that is available to it much more efficiently. The red blood cells pick up and carry more oxygen from the air to the rest of your body.

Doctors find that most people born on Mars have more red blood cells than do people born on Earth. If you ever go on a trip to Earth, though, you'll find that the number of your red blood cells soon drops. The change is like moving from a higher to a lower altitude on Earth.

So, the weaker gravity on Mars is both harmful and helpful. It hurts you because it allows your muscles to become weak and flabby. But, at the same time, your body works more efficiently under low-gravity conditions. By exercising daily you can enjoy all the good effects without suffering from the bad.

5

THE AIR YOU BREATHE

IF YOU LIVED ON MARS . . . you'd be taught from the time you're a tot never to wander outside the colony. Never! Otherwise, you would surely die.

Your lungs are normally filled with air under high pressure. But the air pressure on Mars is extremely low. If you ever leave the colony without a space suit, the low pressure bursts the air out of your lungs and you can't breathe. Death usually follows within about five seconds.

The air pressure on Mars is lower than on Earth because the pull of its gravity is weaker than Earth's gravity. The weaker Mars gravity does not hold as much air to its surface. Less air means lower air pressure.

As a matter of fact, the air pressure at the surface of Mars is very low—only 0.008 pounds per square inch. That's only a tiny fraction of Earth's pressure, which is 14.7 pounds per square inch.

But suppose you could manage to breathe under such conditions. What would happen then?

You still would not survive. The air on Mars contains different gases than does the air on Earth. On Earth, about 21 percent of the air is oxygen, which humans need in order to breathe. Nitrogen makes up about 78 percent. The rest consists of tiny amounts of carbon dioxide, argon, and other gases.

By comparison, the air on the planet Mars is mainly carbon dioxide—about 95 percent. The remainder is mostly nitrogen, with just small amounts of oxygen, argon, and some other gases. It is just not possible for humans to live for more than a few minutes with so little oxygen.

You may be wondering: With very low air pressure and not enough oxygen, how can anyone live on Mars?

The humans actually create their own atmosphere. Engineers control both the gases in the air and the air pressure inside the colony. They make sure that the air contains the same gases and is at the same pressure as on Earth. And the colony is sealed off from the rest of Mars's atmosphere. Thus, the inside air doesn't seep out, and outside atmosphere doesn't seep in.

The air for the colony is prepared in the Life Support Building. This gigantic, windowless building is located on the outskirts of the colony. Upon entering the building, a blast of searing-hot air strikes your face and a thundering roar practically deafens you. As you walk between the mammoth machines you can feel their vibrations in your bones. Your skin tingles from the heat.

Oxygen for the entire colony comes from here. One source is the soil, which contains about 42 percent oxygen. Huge amounts of soil, including lots of big rocks, go into those giant noisy machines. The machines grind everything down into a fine powder. Then the powder is baked at a very high heat.

The heat causes the oxygen to come off as a gas. Pipes carry the oxygen from the tops of the machines to giant gas storage tanks and then to the colony.

The other source of oxygen is the Martian air itself. The air contains very little oxygen, only about one-tenth of 1 percent. But about 95 percent of this air, you recall, is carbon dioxide. Every molecule of carbon dioxide contains one atom of carbon and two atoms of oxygen.

Pumps suck in huge quantities of Mars's air and send it over high-voltage electrodes. The powerful electric current pulls apart the molecules, separating the carbon from the oxygen. The pure oxygen then flows through pipes to storage tanks.

There are also giant air compressors in the Life Support Building. They raise the pressure of the air inside the colony to the level found on Earth.

Although the air content on Mars is continuously monitored and adjusted, the carbon dioxide level rises from time to time. Carbon dioxide affects your vocal cords in a strange way. It makes your voice lower and deeper in sound and gives it a sort of growly quality.

The air pressure, too, is constantly checked by automatic measuring devices. The pressure nearly always stays at the proper level. But every once in a while something goes wrong. The pressure falls. When this happens, people with heart disease or with breathing problems get sick. Such individuals usually have portable oxygen masks to use in case of an emergency.

The Life Support Building also heats, filters, and recirculates the air. This is to make sure that the air is fresh and comfortable all the time. The Life Support System is the heart of the colony. It must be working well every minute of every day or it might spell the end of the colony.

An eight-page color insert follows on pages 27–34.
Chapter 6 begins on page 36.

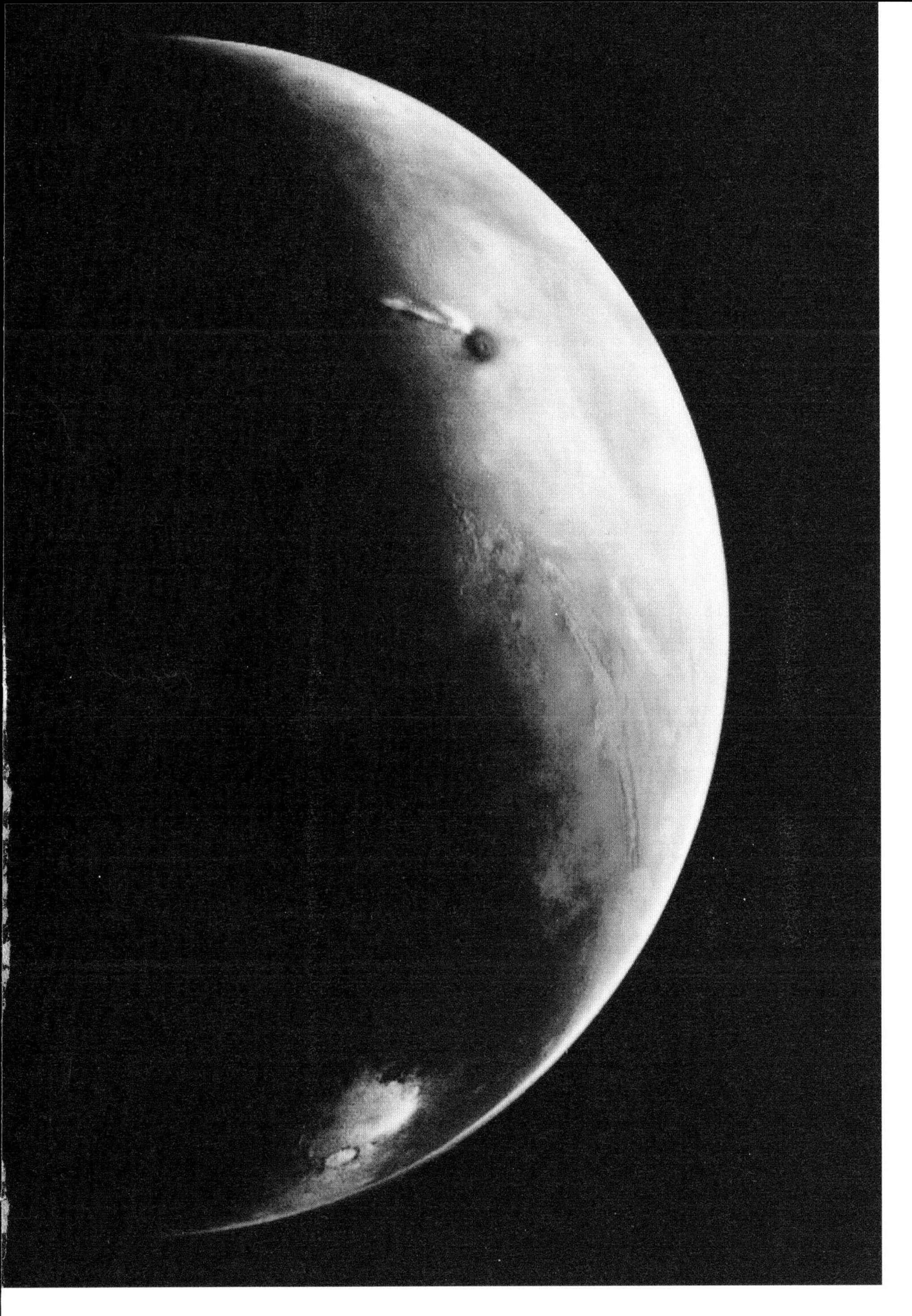

Viking 2 took this picture of Mars from a distance of 261,000 miles.

The surface of Mars is strewn with rocks.

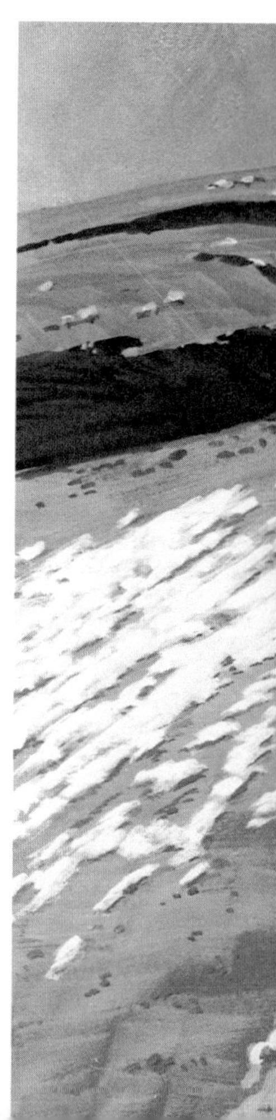

An unmanned vehicle lands on hazardous Martian terrain.

A landing craft awaits the return of astronauts.
CARY HENRIE

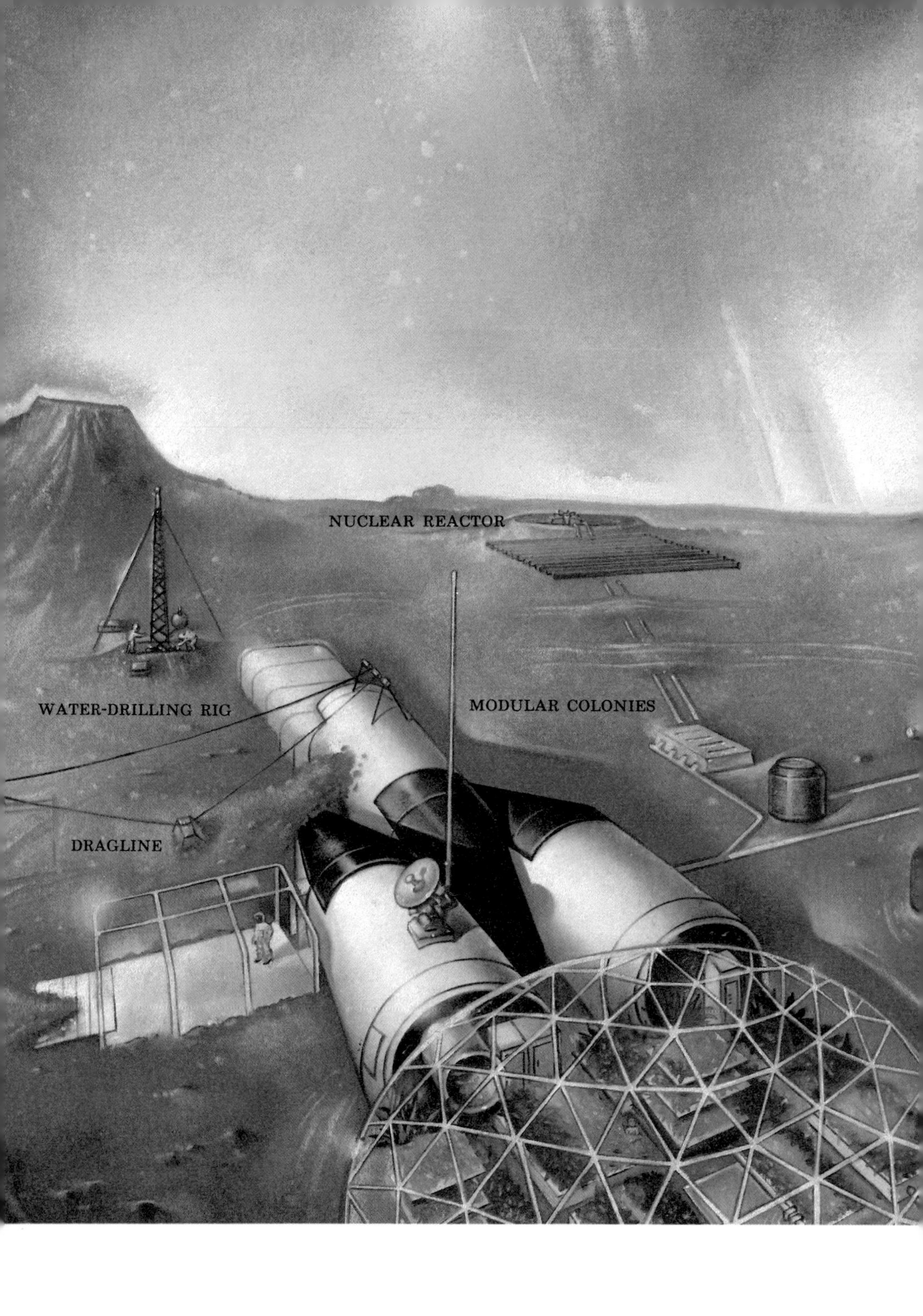
NUCLEAR REACTOR
WATER-DRILLING RIG
MODULAR COLONIES
DRAGLINE

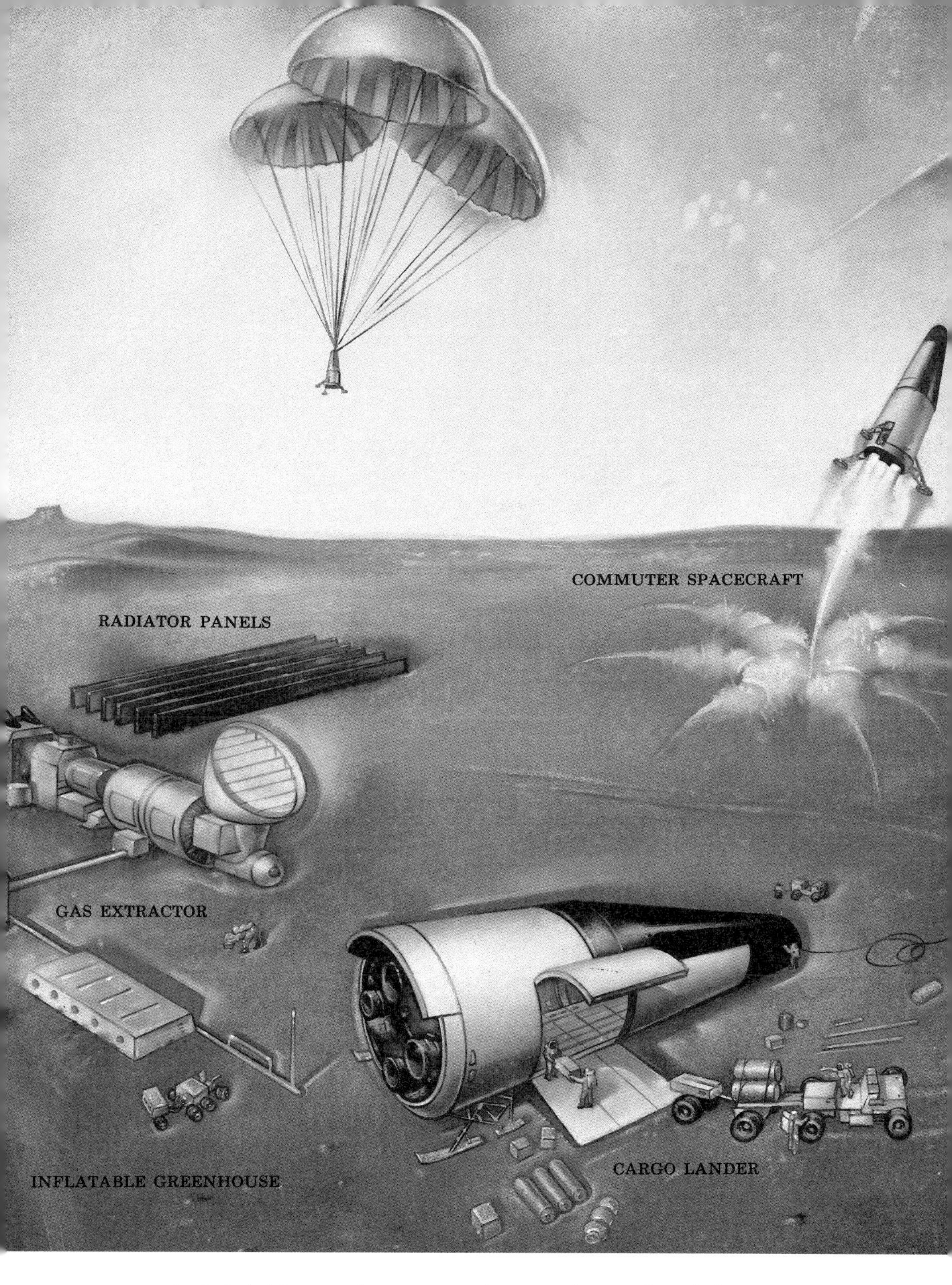

An inflatable greenhouse, a nuclear reactor, and modular colonies, among other things, make living on Mars possible. CARY HENRIE

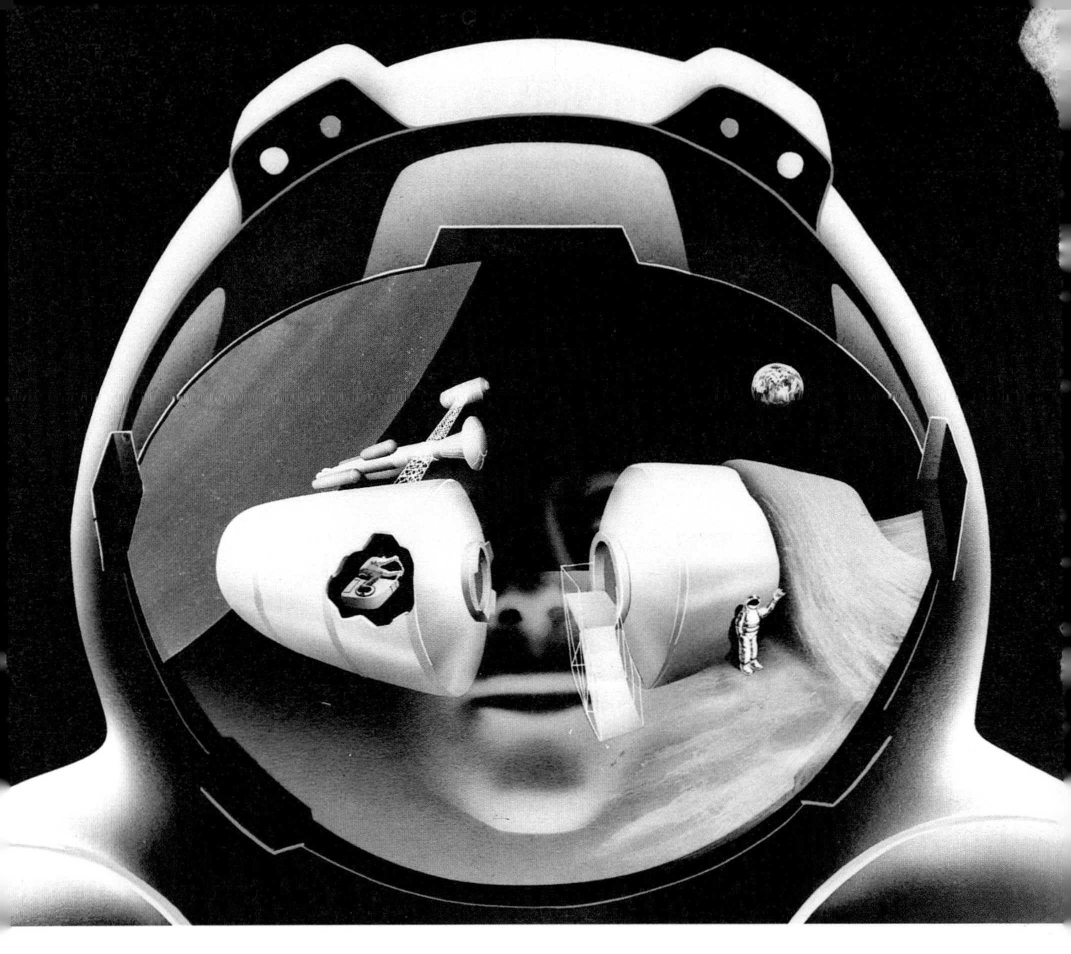

Scientists are researching ways to make living on Mars safe. Here, an extra-vehicular activity suit and a human performance study are reflected on the helmet of a closed-loop life-support system.

THE WATER YOU DRINK

IF YOU LIVED ON MARS . . . you would know that Mars is drier than the driest desert on Earth. No rivers, oceans, ponds, or brooks exist anywhere on Mars.

Yet it is very likely that Mars once had flowing water. Long, long ago, rivers cut several deep canyons and long channels into the soil. There is even evidence that there once was widespread flooding on Mars.

But there came a time in Mars's history when the climate changed. All the water froze. This marked the start of an Ice Age on Mars that is still going on. The present Ice Age can be compared to the one that occurred on Earth about two million years ago.

Now water exists on Mars in only two forms: frozen water, or ice; and water vapor, a gas found in air. Not even one drop of liquid water exists naturally on the planet.

Of course, water is basic to human life. You could

not live more than a few days without water. Therefore, keeping the colonists supplied with water is a major concern.

The very first settlers thought of carrying water to Mars from Earth in rocket ships. The idea did not work because water is so heavy. Someone figured out that water brought up this way would cost about ten thousand dollars an ounce!

Clearly, a better way had to be found. Large deposits of ice—about 800 billion tons—lay frozen in the soil of Mars. Scientists knew that this ice, called permafrost, would be a good source of water. But how to remove the water from the permafrost?

The method they chose is to dig the permafrost out of the ground in big, solid blocks. Because permafrost is harder than many types of rock, this is perhaps the toughest part of the whole process. Drills with diamond tips are used to separate and break apart the permafrost.

The pieces of permafrost are then fed into huge ovens, which are heated by solar energy. The heat melts the ice that is frozen in with the soil and produces liquid water. The water then collects in pipes and passes into big storage tanks.

At this stage, the water from the melted permafrost is not very pure. The many dissolved minerals give it a terribly salty taste and make it too hard for lathering with soap. So workers put it through a chemical process to remove the salts. The result is water that is fit to drink and use for washing—but very costly.

Water vapor, the other source of water, is found in the clouds and fog often seen on Mars. But to get the water from the air is almost as difficult as taking it from ice deposits in the ground. Still, such water tastes far better than the water that comes from permafrost.

To get water from water vapor, vast amounts of Martian air are pulled into another part of the Life Support Building. Here the air is compressed, or squeezed. Next, it is quickly cooled. This causes the water vapor to change into liquid water. Finally, the water drips into pipes and is pumped into tanks where it is kept until needed.

Even with these two sources, water is scarce. Everyone uses as little water as possible. Most bathrooms have showers, not bathtubs, to cut water consumption. Leaky faucets are unheard of. Chemicals are used instead of water in toilets. And industry, which needs quantities of water for jobs like making concrete, is very economical in its water use.

Scientists on Mars are now beginning to dream of rivers flowing with water. They imagine lakes and perhaps a small ocean. When that happens there may even be swimming pools and Jacuzzis on Mars!

7

THE FOOD YOU EAT

IF YOU LIVED ON MARS . . . your diet would be much more limited than that of people on Earth. It would be made up mostly of vegetables and grains, with little meat or fowl and no fish.

The first settlers knew that they could not grow anything in the soil of Mars. For one thing, the ground is too cold. For another, not enough liquid water is available. And finally, the soil does not have the minerals that most plants need to grow.

In the beginning, the settlers brought their food from Earth. Mostly it was freeze-dried, concentrated, or preserved—just like the food the astronauts eat. Food in these forms has several advantages: It doesn't spoil, doesn't weigh much, and doesn't take up much space. But it has one major drawback: It tastes awful!

So the early colonists decided to raise some of their own crops. In their favor was the fact that plants need carbon dioxide from the air to grow. And the Martian air is almost all carbon dioxide. But three problems still remained: keeping the plants warm, supplying them with water, and getting good growing soil.

The solution was obvious: Put up a greenhouse. In short order they built one—with a dome of clear plastic held up by air pressure. They set the greenhouse just beyond the covered part of the colony but joined to it.

The sun heated the greenhouse and kept it at a good temperature for growing plants. Because it was a closed system, all under the dome, the same small amount of water could be used over and over again. And by washing and treating the ordinary Martian soil and adding special fertilizers, they were able to use that dirt for the planting.

One big plus in growing plants on Mars is that there are no bugs or small animals to harm the crops. There were no such pests on Mars before the settlers came, and the authorities keep it that way. They carefully check all new arrivals and shipments of goods to be sure they don't bring any insects or rodents with them. Another advantage of raising crops on Mars is the weak gravity. Less pull allows the plants to grow taller and to bear more than on Earth.

Having seen that the first greenhouse worked, the colonists began building new ones. By now, there are seven huge greenhouses, each the size of two football fields.

Going into a greenhouse is like entering a steam bath. Inside the hot and steamy room you notice plants growing tightly packed together. This approach is called high-density farming. By planting the seedlings one right next to the other, using carefully prepared soil, and controlling the growing conditions, farmers on Mars can produce more food per square foot than do most farmers on Earth.

Two greenhouses are used to grow peanuts and soybeans. Both are excellent sources of protein. The soybean is especially valuable in making a delicious hamburger substitute. And both are pressed to remove their oils, which are then used in dozens of different food products, from margarine to chocolate brownies.

A couple of greenhouses produce fruits and vegetables. The favorite fruit on Mars is the strawberry. Not only do strawberries taste good, but they're small in size, easy to grow, and rich in vitamins. Also found throughout these greenhouses are such vegetable crops as potatoes, beets, broccoli, corn, peas, lettuce, carrots, and tomatoes.

Several greenhouses supply the vital grains—wheat and rice—used in the colony. One building has just wheat growing over its entire area. It looks like a plain in America's Midwest. The other is one big rice paddy, completely covered with water.

The most advanced type of soilless farming, hydroponics, goes on in the last and newest greenhouse. Instead of being set in soil, the roots of the plants are placed in troughs of water to which special fertilizers have been added. Hydroponics allows bigger, healthier plants to grow more quickly and in a smaller space than usual.

Just beyond the greenhouses are two buildings that look—and smell—like regular barnyards. It is here that the small number of animals that are raised on Mars are kept.

The colonists breed mostly rabbits and chickens. They chose these kinds of animals because they are small and can live on garden scraps and feed made from grain by-products. Their flesh is low in fat, tender, and tasty. And, of course, each one provides an added bonus: The chickens lay eggs, and the rabbits multiply *very* fast.

The favorite milk- and meat-giving animals on Mars are goats. Goats give less milk than do cows, but they are smaller, do not eat as much, and are easier to handle. Recently, the farmers have started raising a small herd of larger farm animals—cows, pigs, and sheep. These animals reach a size larger than animals on Earth very quickly because of the low gravity. But they take up so much space and need so much food that their numbers are always kept low.

The animals are grown, in part, for their meat. Some people yearn for the beef, lamb, or pork they ate on Earth. But, as you know, hamburgers, steaks, chops, spare ribs, and bacon are rare treats on Mars. There is simply not enough meat to go around. The animals also provide Martians with a small amount of wool and leather.

The colonists still get part of their food supply from Earth in freeze-dried, concentrated, or preserved form. But they are producing more and more of their own food in these amazing greenhouses and barnyards. Before very much longer all the food needed for a complete, varied, and nutritious diet will be grown right here on Mars.

IT'S A MATTER OF TIME

IF YOU LIVED ON MARS . . . you would use clocks and watches set to Martian time instead of the ordinary ones used on Earth. The reason is that a day on Mars lasts longer than one on Earth.

The length of a day on any planet depends on how long it takes for that planet to spin around itself once, or to make a full rotation.

Mars rotates more slowly than Earth does. Suppose, for example, you were standing on the Earth's equator. You'd be spinning around at 1,044 miles per hour. And it would be exactly 24 hours before you returned to the point from which you started.

But standing on the Martian equator you'd be spinning at only 936 miles per hour. At that speed it would take you 24 hours and 37 minutes to get back to your starting point.

When the original colonists came to Mars, they decided to consider a day on Mars to be 24 hours long, just as it is on Earth, even though it lasts 2.57 percent longer. To keep the 24-hour day, they then had to lengthen the hours, minutes, and seconds by the same 2.57 percent.

Now you know why you need special clocks and watches on Mars. They look like the clocks and watches on Earth. But they all run slightly slower.

But it is not only the days, hours, minutes, and seconds that last longer on Mars. So do the years and seasons.

The length of the year on any particular planet depends on two factors. One is the speed of the planet as it travels in its orbit around the Sun. The other is the length of the orbit.

Earth moves at an average speed of 66,500 miles per hour as it circles the Sun. Mars moves much more slowly, at a speed of only 54,000 miles per hour. The slower speed adds to the length of the Martian year.

Also, Mars, as you know, is about 50 million miles farther away from the Sun than Earth is. That means it has a longer journey around the Sun. In figures, the Earth orbit measures about 584 million miles. The Mars orbit, though, is a full 890 million miles.

Because of the slower speed and longer orbit, the Mars year lasts 687 days. That is far longer than the Earth year of 365¼ days.

Living on Mars, you know that

- you have a birthday only once every 687 days.
- you start first grade when you are three and a half years old.

- at age seven you enter junior high.
- a year and a half later, you begin high school.
- you get your high school diploma when you are ten!

Because a Mars year is longer than an Earth year, each of the four seasons has more days. Further, the seasons vary much more in length on Mars than on Earth.

The number of days in each season depends on the

Because it is farther from the Sun, Mars has a longer orbit. A year on Mars takes 687 days.

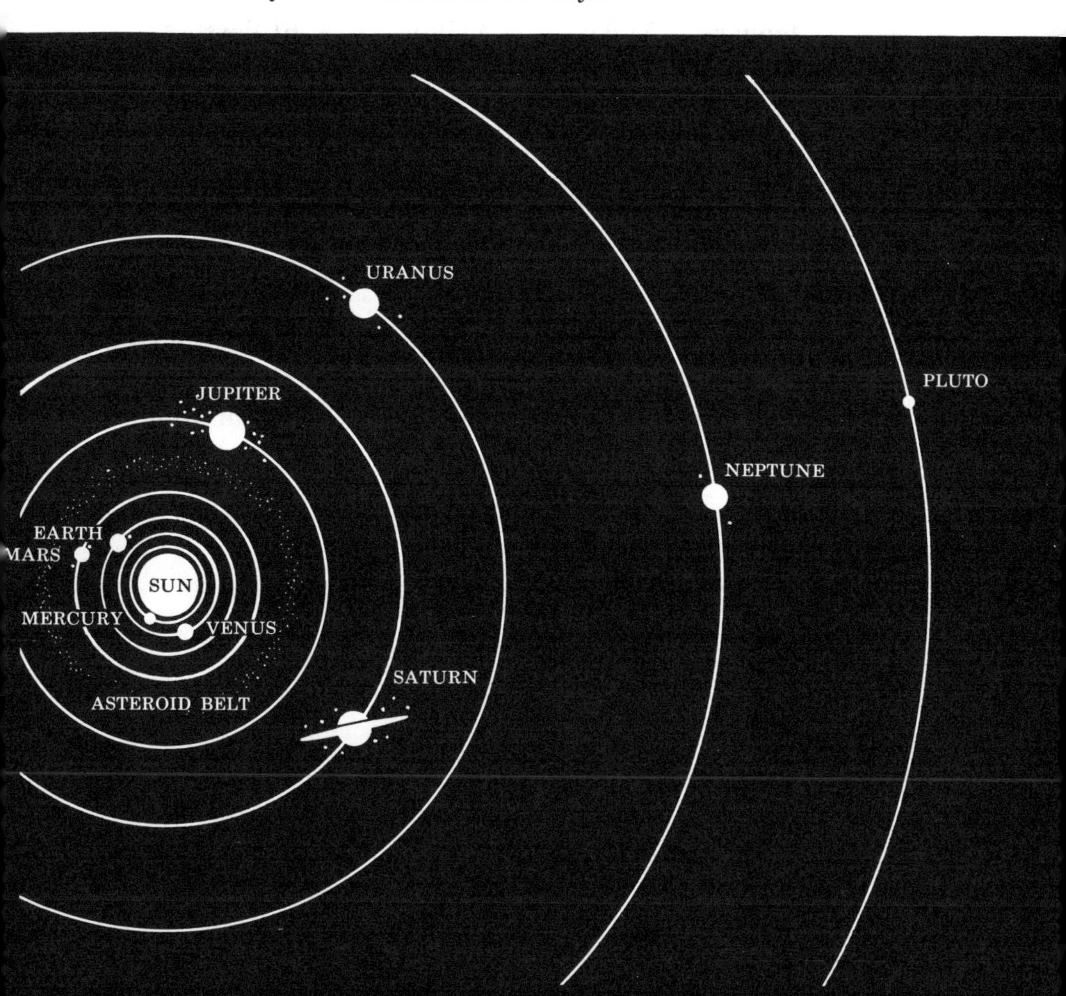

shape of a planet's orbit. Earth's orbit is almost circular, and the seasons are nearly equal in length. Each season lasts between 90 and 92 days.

But the orbit of Mars is shaped more like an ellipse or oval. That brings about big variations in the seasons. On the northern hemisphere of Mars, for example, spring has 199 days, summer 182 days, autumn 146 days, and winter 160 days.

The early colonists adapted their Earth calendars to the longer year and seasons on Mars. The year on Mars has the same twelve months as on Earth. But each month has many more days:

January	53 days	July	61 days
February	53 days	August	60 days
March	66 days	September	49 days
April	66 days	October	49 days
May	67 days	November	48 days
June	61 days	December	54 days

Many of the dates on this calendar seem strange to people used to the Earth calendar. For example, your school year starts around September 7 and ends about June 58. Christmas is on December 46. And your winter recess ends on January 32.

WEATHER—OR NOT

IF YOU LIVED ON MARS . . . you'd never have to worry about the weather as long as you stayed inside the colony.

The temperature and humidity within the colony are always kept at a comfortable level. The climate is as fine as the scientists can make it.

But outside the colony, it is a different story. No rain or snow falls on the surface of Mars. And there are no floods, thunderstorms, or earthquakes (or marsquakes) to trouble you—only occasional dust storms. But the weather is almost always cold. The average temperature is a frigid 100 degrees below zero Fahrenheit.

The low temperatures are caused by the immense distance between Mars and the Sun, the main source of heat in our solar system. Mars is an average of 141.6 million miles distant from the Sun. Compare this to the position of the Earth, only 93 million miles away from the Sun on average. The greater space between Mars and the Sun means that Mars gets much less heat. The effect is the same as that of a fire. The closer you are, the more heat you feel; the farther away, the less heat reaches you.

Summer on Mars is about as cold as winter in northern Canada. For example, on a sunny summer day at noon on Mars's equator, the temperature may get as high as 40 or 50° F.

In the winter the average reading slides down to −140° F. The temperature is even lower around the north and south poles. These are the coldest spots on the planet, with the temperature dipping to −220° F.

On Earth the range between winter and summer temperatures is usually no more than about thirty or forty degrees F. On Mars the change from hottest to coldest days covers a range of more than one hundred degrees F.

This wide gap in temperature is due in part to the shape of Mars's orbit—an ellipse. It makes for a big difference between the closest and farthest points from the Sun.

At its closest (perihelion), Mars is about 128.3 million miles away from the Sun, and it receives an amount of heat about half of what the Earth receives at its perihelion. At its farthest (aphelion), Mars swings out to 154.7 million miles. At this point, Mars gets a bit more than one-third as much heat as the Earth.

Because the shape of the Earth's orbit is much closer to a circle, its perihelion is 91.3 million miles and its aphelion is nearly the same, 94.5 million miles. The slight difference of about 3 million miles keeps the temperatures on Earth within a narrower range.

The tilt of the planets also affects their temperature range. Both Earth and Mars are slightly tilted in relation to the Sun. Earth has a tilt of 23.45 degrees. Mars is tilted just a bit more—25.17 degrees. This tiny extra tilt also contributes to the wide variation in temperature on Mars from season to season.

The colonists cannot withstand the brutally cold temperatures and the lack of oxygen on Mars. You can only survive in two ways: by staying within the colony—where the weather and climate are controlled—or by wearing space suits outside—where you carry a good climate with you.

10

OUTSIDE THE COLONY

IF YOU LIVED ON MARS . . . you'd have to wear a space suit every time you went outside the colony.

The airtight space suit, with its connected helmet and boots, has everything you need to keep you alive. In the backpack are flasks of oxygen under pressure to provide the air you must have to breathe. The backpack also holds the batteries that heat the suit and keep you warm. And a special reflective coating protects you from ultraviolet rays and other kinds of dangerous radiation.

To leave the colony, you have to pass through one of the air locks that lead to the outside. These double sets of doors make sure that there is very little mixing of the outside air with that of the colony.

After going through the first door of the air lock you are in a small room with a second door on the opposite side. Only after you close the first door, sealing in the colony's air, can you open the door that leads to the outside.

Outside the colony, you are struck by how red everything looks—the soil, the rocks, even the sky. It is this overall bloody color that gives Mars its nickname, the Red Planet.

The soil of Mars is red because it contains so much iron. Iron and its compounds make up 12 percent of the soil on Mars, which is more than twice the iron content in the soil on Earth.

Most of the iron in the Martian soil is chemically joined to oxygen. The scientific name for this particular iron compound is either *limonite* or *magnetite.* The common name is *rust.*

The chemical formula for limonite is Fe_30_4. It tells you that each molecule consists of three atoms of iron (the chemical symbol for iron is Fe) joined to four atoms of oxygen (the chemical symbol for oxygen is O).

Now that you know why the soil is red, can you figure out why the sky is red? Why isn't it blue?

The sky is red for the same reason the soil is red—iron. Winds are always blowing on Mars. The constant winds keep swirling tiny grains of the reddish soil around in the air. These red dust particles paint the air—and the sky—red.

On Mars you have to watch out for dust storms. Winds have an average speed of between 15 and 20 miles per hour; the gusts sometimes top 100 miles per hour. Occasionally the wind blows so hard that it amounts to a hurricanelike storm. Such storms can carry ten times as much dust as the worst desert storms on Earth.

Dust storms can sometimes get so bad that they cover the entire planet. Towering columns of dust kicked up by the storm rise to 3.5 miles above the ground. The colonists call these dust storms dust devils.

Dust devils are like tornadoes. Inside the swirls of the dust devil, the dust grains are swept around at greater and greater speeds. Its twisting, swirling winds can pick up objects weighing several pounds and fling them about. The towering cloud of dust stays in one place above the ground or moves at high speeds with the winds.

Scientists studying dust devils have found that sometimes the sun heats the soil very rapidly. If the temperature is right and there are enough loose surface grains, a dust devil is born.

Except for the wind, nothing moves on Mars. All is desertlike—parched and barren on all sides. No water. No trees. No plants or animals. Just sand, rocks, and wilderness from pole to pole.

But one thing may surprise you. Here and there, you may see huge patches of very dark-colored sand. Astronomers on Earth used to gaze at these dark areas on Mars with wonder. Most thought the dark spots were cities or oases. Of course, they are neither. In actual fact, the patches are nothing more than spots of different-colored sand.

The reason for the spots becomes clear as you walk across the Martian landscape. Although the wind blows all the time, it does not blow the same everywhere. Where the winds are strongest, they blow away the light-colored top layer. That exposes the darker

Mars is a desert. Its surface is covered with sand and rocks.

soil underneath; it looks dark red or black. Where the winds don't blow as hard, the lighter topsoil stays in place. These areas are lighter in color.

The astronomers on Earth thought they saw a whole civilization on Mars. But you know better. What they were looking at was simply different levels of Martian soil.

11

OF MARS ROVERS AND AIRSHIPS

IF YOU LIVED ON MARS . . . you would use a Mars rover or an airship for all the trips you took outside the colony.

The Mars rover is a six-wheel-drive vehicle with huge balloon tires. It looks like an armored truck—big and sturdy—but with lots of windows. Built in are oxygen tanks and an air compressor. With the doors closed, the rover is airtight to maintain a healthy air supply for everyone inside. The engine is operated by electricity. The power comes from two sources—batteries and the solar panels on the roof.

At the wheel of every rover is a driver who is very familiar with the Mars terrain. He or she is also specially trained to repair the vehicle and handle any emergency.

Before leaving, the driver may ask you to help get ready. Together you check the batteries and solar cells and make sure the engine is running smoothly. The two of you also load all the necessary food, water, and other supplies into the rover. You have to pack everything you need, plus plenty of extras in case of an accident or emergency. After all, you can't pull into a service station to get a flat tire fixed or drive to a supermarket if you run out of bread!

The air in the rover is carefully regulated, so you ride in street clothes. But if you want to step outside, even for an instant, you must first put on a space suit.

Because it carries so many supplies, the rover can only be on the road for a few days at a time. And because of the rough terrain, it can cover no more than 50 or 60 miles a day.

Driving on Mars is not easy—even in a Mars rover. The surface is sandy and icy. If you're not careful the tires can skid and slip. Giant stones and rocks are another menace. They are scattered everywhere.

While in the rover you keep in constant radio contact with the colony. Radio messages are beamed up to one of the communications satellites in orbit around the planet. From the satellite, the signal flashes down to the colony. The messages from the colony return to the rover the same way.

Radio signals cannot be broadcast over long distances as they can on Earth. Earth's atmosphere reflects back radio waves so they can travel across hundreds or thousands of miles. Mars's atmosphere does not reflect radio waves. All radio communication outside the colony, therefore, takes place via satellites.

You use land rovers for short trips. For covering greater distances, you depend on airships. An airship

The view from the Mars rover is of rocks and stones on all sides.

is a cross between a balloon and an airplane. The balloonlike part is a giant, airtight bag that is supported on the inside by a thin metal frame. The bag is filled with hydrogen gas, which is lighter than the Mars air. This causes the airship to float above the surface. Since the air on Mars has so little oxygen, there is very little danger that the hydrogen will catch fire or explode.

The airplanelike part is a long, narrow, pressurized cabin for the passengers and crew suspended beneath the bag. Behind the cabin are two propellers that move the craft forward. Like the rover's engine, the electric engine for the propellers gets its power from batteries and solar cells.

You board the airship at the colony's spaceport. (This is also where the rockets from Earth arrive.) From the inside, the spaceport looks like a small airport on Earth. But it is not exactly the same. There are no runways, since neither rockets nor airships need them for launches or landings. And all the craft are docked at a landing port.

The landing port itself looks like a huge, narrow garage that is open at one end. The rocket or airship gently glides in through the opening. Powerful mechanical arms from the sides and bottom of the port lock the craft firmly in place. A spaceport attendant operates the controls that move a long, covered walkway to the door of the craft. Here, another worker joins them with an airtight seal. The passengers and crew enter and leave through the walkway.

On your first airship flight, you brought along a compass someone had sent you from Earth. You wanted to chart the course of the flight. But the compass didn't work. The needle just wiggled back and forth.

Later you learned that a compass can work only if there is a lot of iron in the core of the planet, as on Earth. Martian soil has plenty of iron. But its core contains very little iron. Therefore, compasses can't be used here. To navigate the airship, the pilot depends completely on radio signals from satellites.

Airship flights can last several days. And they can take you to some of the most fascinating spots on Mars.

12

SEEING THE SIGHTS

IF YOU LIVED ON MARS . . . you would be able to see the most amazing features of the Mars landscape from an airship.

All of your trips start from the colony, which is in the northern hemisphere of the planet. (The northern hemisphere was chosen simply because it is more like Earth than is the southern hemisphere.) Flying over this area, you see lots of volcanoes and big areas covered with lava. Huge cracks and breaks pockmark the surface. But more than anything else, the land resembles a rough, rocky, red desert.

The most striking sight you see in the northern hemisphere is the Great Tharsis Bulge. You cannot miss it; it is a giant area that rises high above the surface.

Located here is a chain of four mammoth volcanoes. The biggest one is called Mount Olympus or

The four huge volcanoes of the Great Tharsis Bulge as seen from a height of 250,000 miles. Mount Olympus is at the bottom on the right.

Olympus Mons. It is the largest mountain on Mars. But what makes it even more special is that it may be the highest peak in the entire solar system.

Mount Olympus rises a full 15 miles above the ground. At its summit is a 50-mile-wide basin, or caldera. And its base is 375 miles across. This one mountain is actually bigger than the entire state of Missouri!

Compare Mount Olympus with the largest volcano on Earth, Mauna Loa in Hawaii. A mere 5.5 miles high, Mauna Loa's base on the bottom of the Pacific Ocean is just 124 miles wide.

The three other volcanoes in the Great Tharsis Bulge are each more than 10 miles high. At one time they were simply called South Spot, Middle Spot, and North Spot. Later, they were given Latin names: Arsia Mons, Pavonis Mons, and Ascraeus Mons.

From the air, the southern hemisphere looks like the Moon. Hundreds and hundreds of craters—some big, some small—dot the surface. What caused them? Probably meteorites crashing into the planet. Unlike the northern hemisphere, the southern hemisphere has no signs of volcanoes.

The most fantastic sight in the southern hemisphere is Valles Marineris. It is an immense east–west gash in the surface. The name comes from the Mariner space shots which sent back the first picture of this land formation. Some people call Valles Marineris the Grand Canyon of Mars. As a matter of fact, though, you could fit a few Grand Canyons in there. The Valles Marineris is about 10 times longer, 40 times wider, and twice as deep as the Grand Canyon!

The Valles Marineris stretches for over 4,000 miles. Its width from rim to rim varies from 30 to 800 miles. And while its average depth is about 1.5 miles, in places it is almost 4 miles deep.

Some experts say that the Valles Marineris began as a fracture, or split, in the surface of Mars. Then, over the centuries, it grew wider and wider. Others hold that it was formed like the Grand Canyon. That is, a long time ago a big, powerful river—like the Colorado—flowed here. It dug its way down through the surface to form the canyon. Scientists keep study-

The upper part of this photo shows the craters on the surface of Mars. The lower part shows the immense Valles Marineris.

ing the soil and rocks in the sides of the Valles Marineris. But they're still not sure exactly what happened.

A special feature of some airship flights is a stop at the ice cap at either the north or south pole of Mars. The two poles are covered with layers of ice that are between 3 feet and one-half mile thick. According to one theory, much of the water that once flowed on Mars is now frozen in these giant polar ice caps. Together they extend over more than 4 million square miles of Mars around the two poles.

The caps do not stay the same size, however. During the summer months, they shrink and almost disappear. During the winter they grow bigger.

At first everyone thought the polar caps were just ice, or frozen water. Then they learned that the caps also contain a great deal of frozen carbon dioxide, or dry ice. The caps are made up of many, many thin separate layers of ice and dry ice.

On your flight to Mars's north pole, the pilot lands the airship at the rim of the ice cap. You are able to go out—in your space suit—and walk around. The

The edge of the north polar ice cap in October. The cap is smaller in size during the warm months.

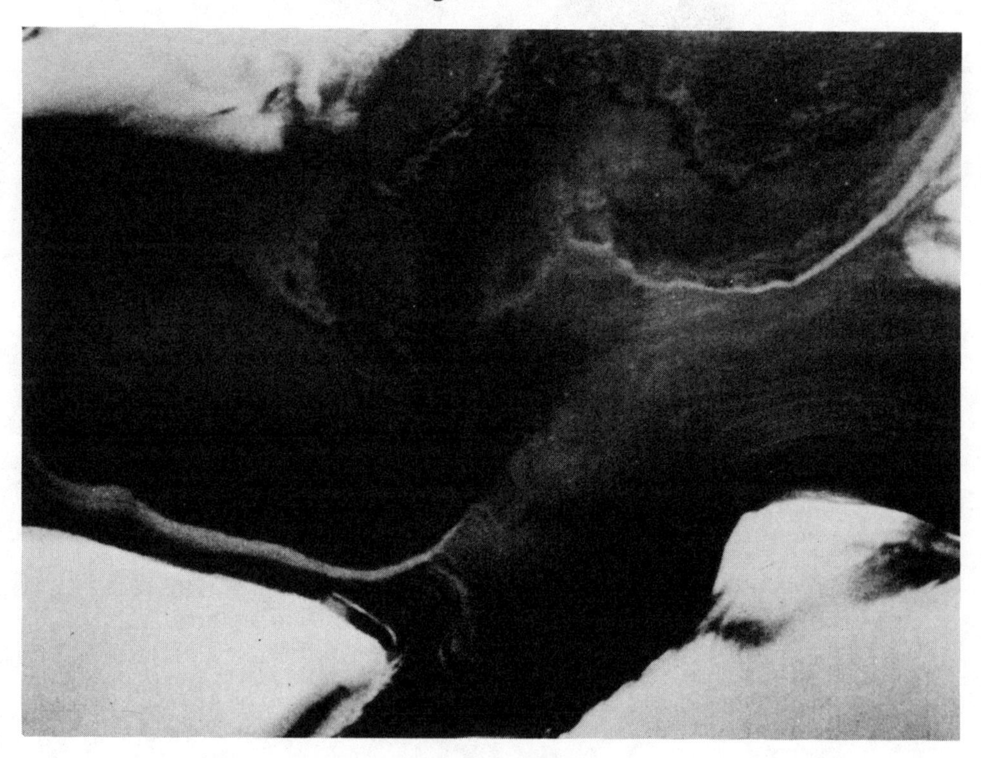

pilot also lends you a portable heater, like a hair dryer, and tells you to aim it at the edge of the ice. To your amazement, the ice doesn't melt and turn to water. It just vanishes into thin air!

The pilot explains that because of the low atmospheric pressure on Mars, melting ice changes directly into water vapor, not water. And it causes the frozen carbon dioxide to change into carbon dioxide gas. The vapor and the carbon dioxide then mix with the air and disappear.

Even though you are wearing a heated space suit, you are chilled by the frigid air near the north pole. When the pilot says, "Let's go," you're ready to return to the airship's warm cabin for the flight back to the colony.

13

THROUGH THE TELESCOPE

IF YOU LIVED ON MARS . . . you would get a better view of the sky than people on Earth. All the stars and planets would appear sharper and clearer.

Telescope viewing on Mars is better than on Earth because the Martian atmosphere is thinner than the Earth's and interferes less with clear viewing. Also, the sky is much darker and therefore there is more contrast between the light of the stars and the background. This makes it possible to see many more dim objects in space. The darkness of the sky is a result of two facts: The colony is completely covered; no light spills out from there. And Mars does not have a large moon to brighten the sky.

From Mars the Sun looks quite a bit smaller and less bright than it does from Earth. In fact, the Sun appears to be only about two-thirds as big and brilliant because Mars is so much farther away.

You can see the stars and planets more clearly from Mars than from Earth.

Earth, at 50 million miles, is the closest planet to Mars. Through the lens, you mostly see the clouds surrounding Earth and catch glimpses of its blue oceans. Venus, the next nearest planet, is nearly 75 million miles distant. You see it as a small, brightly glowing, yellow-white sphere.

It is hard to make out any details on the other planets, even with the better viewing from Mars. The planets range from Mercury, which is over 100 million miles away, to Pluto, which is at the edge of the solar system, at a distance of more than 3 billion miles.

Perhaps the most striking sight is the two moons

Through a powerful telescope on Mars you can see the clouds surrounding Earth.

that circle Mars. Mars's moons were discovered in 1877 by the American astronomer, Asaph Hall. He named them Phobos and Deimos after the legendary horses that pulled the chariot of Mars, god of war.

Through the telescope, Phobos, the closer moon, looks like a great big potato, rolling and tumbling in space. Phobos is quite small, only about 17 miles in diameter. Compare this to the Earth's Moon, which has a diameter of about 2,160 miles.

But Phobos is much closer to Mars than the Moon is to Earth. The distance to Phobos is only 3,750 miles; the distance from the Earth to its moon is 239,000 miles. Looking through the telescope, Phobos appears one-third the size of the Earth's moon.

You also notice that the surface of Phobos is covered with pits and craters. (One astronomer called it a diseased potato.) Some of the marks on the surface are quite big, nearly 2 miles across.

Once astronomers thought that—like the craters on Earth—the craters on Phobos were formed by volcanoes. But now the experts believe that the craters were caused by huge objects—like meteoroids or asteroids—slamming into it.

Meteoroids and asteroids are found orbiting in immense numbers just beyond Mars, between Mars and Jupiter. Every once in a while, one of them breaks free and smashes into Phobos. In fact, some scientists think that Phobos itself may have been a falling asteroid that went into orbit around Mars.

Phobos revolves around Mars at the incredible speed of about 4,788 miles an hour! That's twice the speed of the Earth's Moon. Phobos makes three complete swings around Mars every single day. Time really flies on Phobos. There are actually three Earth months in one day on Phobos!

But here's the really unusual part. Normally, you would expect Phobos to rise in the east and set in the west. But since Phobos revolves so much faster than Mars, it appears to rise in the west. Then it flashes across the sky the wrong way and sets in the east in about 4.5 hours. And some 11 hours later it rises again.

You have more trouble spotting Deimos, the other moon. It has the same potato shape as Phobos. But it is even smaller—only 10 miles in diameter. And it is much more distant—about 12,000 miles from Mars.

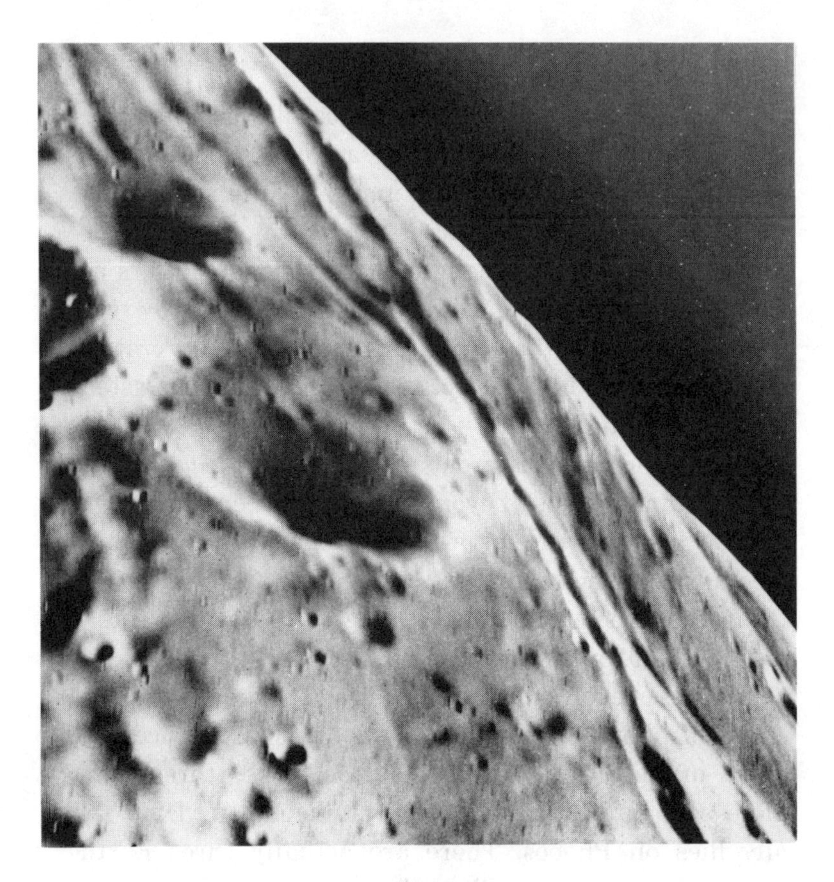

The surface of Phobos is covered with so many pits and craters that it resembles a "diseased potato."

Even with a powerful telescope, you have to strain your eyes to catch sight of Deimos. Because it is so small and so far away, Deimos looks like a dot of light in the sky. Unless you know what you are looking at, you might think you are seeing a very distant star.

Deimos revolves more slowly than Phobos, at only 2,628 miles per hour. Deimos takes a little longer than one Martian day—30 hours and 18 minutes—to make a complete revolution. Therefore it remains in the sky

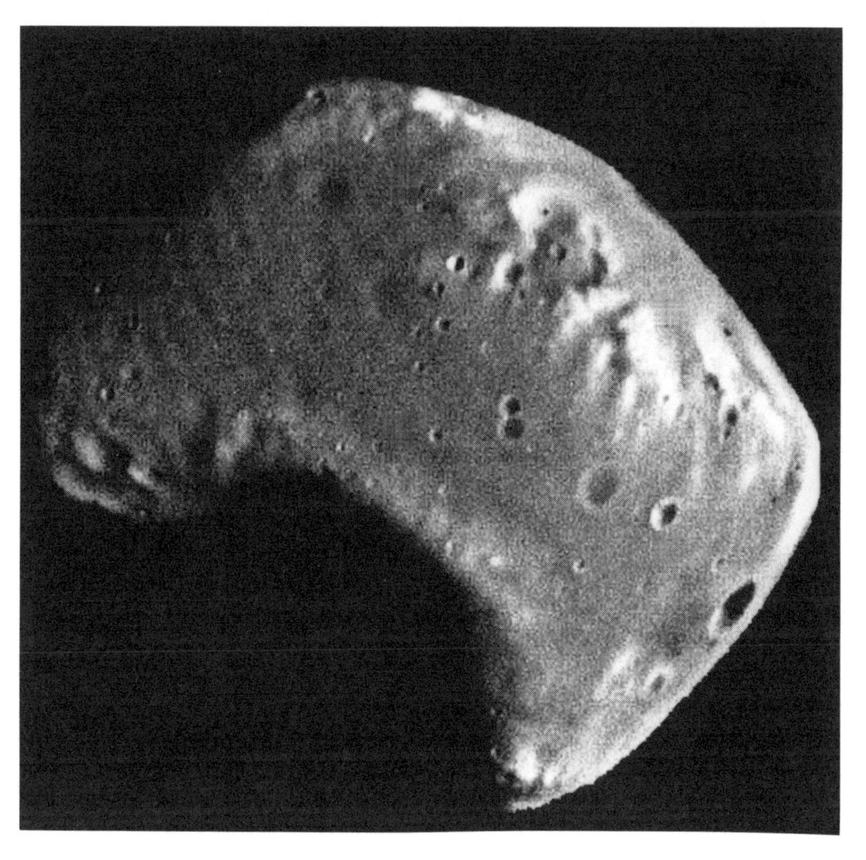

Deimos, the second moon, is small. It is only 10 miles in diameter.

for some 60 hours before it sets. And its surface is smooth and unbroken, nothing at all like Phobos.

Astronauts have landed on Phobos and Deimos many times. But because the moons are so small, they have hardly any gravity. An astronaut who weighs 150 pounds on Earth and 57 pounds on Mars weighs just over 1.5 pounds on Phobos. And when the same astronaut lands on Deimos, he or she weighs less than one ounce!

Phobos and Deimos are natural satellites. But there are also about twenty artificial satellites in orbit around Mars.

Since the Martian atmosphere does not bounce back electronic signals, communication between the colony and a Mars rover or airship must be carried on through a satellite. And, of course, the airships navigate by means of signals the satellites beam down.

With its two natural satellites and twenty artificial satellites, the Martian sky is pretty crowded!

14

THE FUTURE OF MARS

IF YOU LIVED ON MARS . . . you would know about the exciting plans for the future of the Red Planet.

Many people want to live in other areas, not just in the colony. And a group of scientists is working to achieve this goal.

The overall plan they are following is called terraforming. The science fiction writer Jack Williamson created the word way back in 1942. It means changing a planet to become more like Earth. When Williamson first wrote about terraforming it was a wild idea thought up by a writer of fantasy. Today it has become an important subject for serious research in Mars's laboratories.

Scientists expect that there will be room for millions or even billions of people all over Mars. Even though it is half the size of Earth, Mars has the same land area. This is because 70 percent of Earth's surface is covered by oceans.

A trip to the main terraforming lab is an amazing experience. You go first into a room with blinding lights and enormous mirrors. It makes you wish you were wearing your sunglasses.

The scientists here are working on a scheme to use mirrors to reflect and focus the Sun's rays onto the surface of Mars. The idea is to raise the temperature of Mars. They hope that the increased heat will melt the polar caps and the permafrost. This will change the frozen water into water vapor. And the water vapor will become part of the air.

Water vapor in the atmosphere acts like a one-way gate. It lets in the heat of the Sun. But it prevents the heat from Mars's surface from escaping. This is called the greenhouse effect. Scientists will succeed in producing the greenhouse effect if they can raise the surface temperature of Mars by just 1 percent!

One important problem has already been overcome. Mirrors made of glass, a material that is very heavy and breaks easily, have been replaced with mirrors made of Mylar. Mylar is the thin film used in recording tapes. Scientists coat the Mylar with a microscopic layer of a chemical that reflects light. Then they stretch the Mylar over gigantic, lightweight plastic frames. The Mylar mirrors are light, unbreakable, and as big as a half-mile square.

The plan is to place eight such mirrors in orbit, aiming them at the polar ice caps. More will be added as time goes on.

At first very little water vapor will be produced.

But because of the greenhouse effect, more and more water will become available. The climate will get warmer. At the same time, the atmospheric pressure will increase.

Another result will be an increase in the amount of liquid water on Mars. There will be rainshowers and snowfalls. And gradually streams and lakes and even oceans will form on the surface.

The scientists have still another scheme. They want to blow a thin covering of Martian soil over the polar caps. Everyone knows that wearing dark clothes in the summer makes you feel warmer than wearing light colors. That is because dark colors absorb heat. In the same way, the dark Martian soil will soak up the Sun's rays and help to raise the temperature of the polar caps. And this heat rise will boost the effect of the mirrors.

It may take twenty, fifty, or even a hundred years to warm Mars. But the scientists are sure that sooner or later they will have Earthlike conditions on the Red Planet.

In time, Mars will have a comfortable temperature and air pressure and plenty of liquid water. But what about the air? Will it ever be fit to breathe?

Work on this problem goes on in another lab. When you enter, you see several huge glass tanks along the walls. The bottom of each one is covered with a layer of Martian soil. But covering the red soil and the walls of the tanks is something that looks like green slime or mold.

The green slime is made up of blue-green algae. Algae are tiny plants that have no true roots, stems, or leaves. This particular type comes from around the Earth's South Pole. It is one of the few forms of life growing in that region.

A plant from the Earth's South Pole is of special interest to the scientists on Mars. Although the South Pole is in the middle of a giant ice cap, it almost never rains or snows there. And in many ways the very dry and cold conditions resemble those on Mars. Any plant that can grow at the South Pole should be able to grow well on Mars.

The purpose of growing the algae is not to plant beautiful gardens. It is to change the air surrounding the planet. Plants remove carbon dioxide from the air and release oxygen. Right now, there's too much carbon dioxide and not enough oxygen in Mars's air for humans to breathe. As the algae grow and spread, they will change the air in exactly the right way.

In fact, the algae will do even more. They will also change the nitrogen in the air into compounds that are suitable for plant food. At the same time they will produce a gluelike substance to help hold the particles of soil together. That will stop the topsoil from being blown away by the Martian winds. In time, this will tend to create soil rich enough to farm.

With mirrors to warm the planet and algae to change the air and fertilize the soil, it will be possible someday for humans to live anywhere on the Red Planet. Mars will be a green planet, with lush fields, flowers, and trees. And many men, women, and children will enjoy the beauties of this once cold, dry, barren world.

FURTHER READING

Books

Berger, Melvin. *Planets, Stars and Galaxies.* New York: Putnam's, 1978.

French, Bevan M. *Mars: The Viking Discoveries.* Washington, D.C.: NASA, 1977.

Lovelock, James, and Michael Allaby. *The Greening of Mars.* New York: St. Martin's Press, 1984.

McDonough, Thomas R. *Space: The Next 25 Years.* New York: John Wiley & Sons, 1987.

Powers, Robert M. *Mars: Our Future on the Red Planet.* Boston: Houghton Mifflin, 1986.

Magazines

Angier, N. "Humans to Mars? Why Not?" *Time*, July 29, 1985, p. 67.

Carroll, M. W. "The First Colony on Mars." *Astronomy,* June 1985, p. 6.

FURTHER READING

Mann, P. "Commission Sets Goal for Moon, Mars Settlements." *Aviation Week and Space Technology,* March 24, 1986, p. 80.

"Manned Mission to Mars." *Science Digest,* November 1984, p. 40.

"Mission to Mars." *Science Digest,* March 1986, p. 26.

"Next Stop: Mars." *Scientific American,* December 1986, p. 74.

Singer, S. F. "The Case for Going to Mars." *Newsweek,* October 6, 1986, p. 13.

"Special Report: Mars or Bust." *Discover,* September 1984, p. 12.

Wilford, J. N. "Destination: Mars." *New York Times Magazine,* March 20, 1988, p. 20.

INDEX

Page numbers in *italics* refer to illustrations.

INDEX